# A Herd of Deer



ANIMAL GROUPS

IN THE FOREST

*Alex Kuskowski*

CONSULTING EDITOR, DIANE CRAIG, M.A./READING SPECIALIST

A Division of ABDO
ABDO
Publishing Company

**visit us at www.abdopublishing.com**

Published by ABDO Publishing Company, a division of ABDO, P.O. Box 398166, Minneapolis, Minnesota 55439. 

Printed in the United States of America, North Mankato, Minnesota
062012
092012

PRINTED ON RECYCLED PAPER

Editor: Liz Salzmann
Content Developer: Nancy Tuminelly
Cover and Interior Design and Production: Anders Hanson, Mighty Media, Inc.
Photo Credits: Shutterstock

Library of Congress Cataloging-in-Publication Data
Kuskowski, Alex.
A herd of deer : animal groups in the forest / Alex Kuskowski.
p. cm. -- (Animal groups)
ISBN 978-1-61783-539-1
1. Forest animals--Behavior--Juvenile literature. 2. Social behavior in animals--Juvenile literature. I. Title.
QL112.K87 2013
591.73--dc23

2012009030



## SANDCASTLE™ LEVEL: FLUENT

SandCastle™ books are created by a team of professional educators, reading specialists, and content developers around five essential components—phonemic awareness, phonics, vocabulary, text comprehension, and fluency—to assist young readers as they develop reading skills and strategies and increase their general knowledge. All books are written, reviewed, and leveled for guided reading, early reading intervention, and Accelerated Reader® programs for use in shared, guided, and independent reading and writing activities to support a balanced approach to literacy instruction. The SandCastle™ series has four levels that correspond to early literacy development. The levels are provided to help teachers and parents select appropriate books for young readers.

Emerging Readers
*(no flags)*

Beginning Readers
*(1 flag)*

Transitional Readers
*(2 flags)*

Fluent Readers
*(3 flags)*

# Contents

# Animals in the Forest

A forest is a place where many trees and plants grow. Wild animals feel safe there. The trees provide **shelter**. The plants give many animals food to eat.

# Why Live in a Group?

Lots of animals live in groups. Animals in a group can **protect** each other. They can share space, food, and water. They also work together to help raise babies. Many animal groups have fun names!

# A Herd of Deer

Deer **roam** together in a herd. In the winter, a herd can be as large as 30 deer. Their brown coats help them hide in the forest.

## Deer Names

**MALE**
*buck, stag*

**BABY**
*fawn*

**FEMALE**
*doe*

**GROUP**
*herd, mob*

# An Aerie of Eagles

A family of eagles is called an aerie. Eagles build their nests in tall trees. They watch over their nests to keep their young safe.

## Eagle Names

**MALE**
*male*

**FEMALE**
*female*

**BABY**
*fledgling, eaglet*

**GROUP**
*aerie, convocation*

# A Sleuth of Bears

Bears are usually **solitary**. But sometimes bears gather to eat. A sleuth of bears can eat a lot of salmon!

## Bear Names

**MALE**
*boar*

**FEMALE**
*sow*

**BABY**
*cub*

**GROUP**
*sleuth, sloth*

# A Gang of Elk

One elk is **vulnerable** in the forest. A gang of elk is much safer. Elk are always watching for predators.

## Elk Names

**MALE**
*bull*

**FEMALE**
*cow*

**BABY**
*calf*

**GROUP**
*gang, herd*

# A Parliament of Owls

A group of owls is called a parliament. Owls often build nests in trees. They lay their eggs there. They will attack animals that come too close.

## Owl Names

**MALE**
*male*

**FEMALE**
*female*

**BABY**
*owlet, fledgling*

**GROUP**
*parliament, stare*

# A Pack of Wolves

Wolves live in a pack. They take care of each other. They hunt for **prey** in the forest.

Wolf Names

**MALE**
*dog*

**FEMALE**
*bitch*

**BABY**
*pup, whelp*

**GROUP**
*pack, rout*

# A Squad of Squirrels

A group of squirrels is called a squad. Squirrels can live in trees or on the ground. They **whistle** to warn each other of danger.

## Squirrel Names

**MALE**
*buck*

**FEMALE**
*doe*

**BABY**
*pup, kit, kitten*

**GROUP**
*squad, dray*

# More

## FOREST GROUPS

A cete of badgers

A sounder of boars

A down of hares

A covey of partridges

A pride of peafowl

A prickle of porcupines

A gang of weasels

A host of sparrows

# Quiz

1. Forests make animals feel safe. ***True or false?***

2. Eagles build nests in short trees. ***True or false?***

3. Bears eat salmon. ***True or false?***

4. Owls never attack animals that come close to their nests. ***True or false?***

5. Squirrels **whistle** to warn about danger. ***True or false?***

ANSWERS: 1) TRUE 2) FALSE 3) TRUE 4) FALSE 5) TRUE

# Glossary

**prey** – an animal that is hunted or caught for food.

**protect** – to guard someone or something from harm or danger.

**roam** – to wander or walk around.

**shelter** – protection from the weather.

**solitary** – being or living alone.

**vulnerable** – able to be hurt or attacked.

**whistle** – to make a loud, high sound.